Ethological Endocrinology

Photovoltaics, Fuel Cells, And Energy Storage Solutions

Janice Frías

Table of Contents

1 Introduction to Ethological Endocrinology

Harnessing the sun's energy through photovoltaic (PV) cells represents a cornerstone of renewable energy. These cells, typically made of silicon, convert sunlight directly into electricity via the photovoltaic effect. The process involves photons striking a semiconductor material, exciting electrons and creating an electrical current. Different PV technologies exist, each with its advantages and disadvantages. Crystalline silicon cells boast high efficiency but can be expensive, while thin-film technologies like cadmium telluride (CdTe) or copper indium gallium selenide (CIGS) offer lower costs and flexibility, though with slightly lower efficiency. Concentrated solar power (CSP) systems utilize mirrors or lenses to focus sunlight onto a smaller area, increasing the intensity and improving efficiency, often employing thermal energy storage for consistent power delivery. Understanding the diverse PV technologies available is crucial for selecting the most appropriate solution based on factors such as cost, efficiency, and available space.

Fuel cells offer another avenue for clean energy generation. Unlike batteries, which store energy, fuel cells continuously convert chemical energy into electricity as long as fuel and oxidant are supplied. Hydrogen fuel cells, for instance, use hydrogen as fuel and oxygen from the air as the oxidant, producing only water as a byproduct—a remarkably clean process. The electrochemical reaction within the cell generates electricity directly, with high efficiency and low emissions. However, challenges remain in hydrogen production and storage, as efficient and cost-effective methods are still under development. Direct methanol fuel cells (DMFCs) offer an alternative, using methanol as fuel, suitable for portable applications due to the easier handling and storage of liquid methanol compared to hydrogen gas. While DMFCs offer convenience, their efficiency tends to be lower than hydrogen fuel cells, and further research is needed to improve their performance and address concerns about methanol toxicity.

Efficient energy storage is critical for integrating intermittent renewable energy sources like solar and wind power into the grid. Various energy storage mechanisms exist, each with its own characteristics and suitability for different applications. Battery technologies, such as lithium-ion batteries, dominate the market for smaller-scale storage, offering high energy density and relatively fast charging and discharging rates. However, issues related to lifespan, cost, and environmental concerns associated with the mining and recycling of certain materials require ongoing research and development. Capacitors and supercapacitors provide high power density, suitable for applications requiring rapid energy delivery, but their energy density is relatively lower compared to batteries. Large-scale energy storage systems include pumped hydro storage (PHS), a mature technology that utilizes excess energy to pump water uphill and then releases it to generate electricity during peak demand. Thermal energy storage (TES) systems store energy as heat, while compressed air energy storage (CAES) stores energy by compressing air. The choice of storage technology depends on factors such as the scale of the application, required power and energy capacity, and cost considerations.

The effective integration of renewable energy sources into existing energy grids presents significant challenges. Hybrid systems, combining different renewable energy technologies, offer a viable approach to overcome the intermittency of individual sources. Solar and wind power can be combined, for example, to provide a more consistent power supply. However, careful system design and control strategies are essential to manage the fluctuating power output from renewable sources. Grid integration challenges include ensuring grid stability and reliability, managing power fluctuations, and developing smart grids capable of handling the intermittent nature of renewable energy. Advanced grid management systems and energy storage technologies are crucial for seamless integration and optimal utilization of renewable energy sources. Efficient energy distribution networks and smart grid technologies are vital for optimizing the delivery of renewable energy to consumers and minimizing losses.

The pursuit of sustainable energy solutions requires continuous advancements in materials science and manufacturing processes. Nanomaterials offer significant potential for enhancing the performance of renewable energy technologies. For example, nanostructured materials can improve the efficiency of solar cells and fuel cells. Additive manufacturing, also known as 3D printing, provides opportunities for creating customized designs and components with improved performance characteristics. Sustainable material selection is essential for minimizing the environmental impact of renewable energy technologies, focusing on materials that are readily available, recyclable, and have low toxicity. Using sustainable and readily available materials in the manufacturing process of renewable energy technologies is key to reducing the environmental footprint associated with renewable energy technologies production, and making the transition towards truly sustainable energy even more effective.

1.1 The Scope of the Field

The field of renewable energy encompasses a vast and interconnected landscape of technologies, each playing a crucial role in the global transition towards sustainable energy systems. Understanding this interconnectedness is paramount to effectively addressing the multifaceted challenges and opportunities inherent in this rapidly evolving sector. This chapter aims to provide a foundational overview, setting the stage for a deeper exploration of the individual components that constitute this vital area of technological advancement.

From the sun's radiant energy harnessed by photovoltaic cells to the controlled chemical reactions within fuel cells, and the diverse mechanisms employed for energy storage, renewable energy solutions present a complex web of scientific principles and engineering challenges. The interdependency between these technologies is evident; effective energy harvesting requires robust and efficient storage solutions to mitigate the inherent intermittency of sources such as solar and wind power. Consequently, advancements in one area often necessitate improvements in others, driving continuous innovation across the entire spectrum.

Consider, for instance, the advancements in battery technology. The increasing demand for electric vehicles and grid-scale energy storage are driving significant research and development in lithium-ion batteries and beyond, pushing the boundaries of energy density, lifespan, and charging speed. These advancements directly impact the viability and practicality of large-scale solar and wind farms, enabling smoother grid integration and enhanced reliability of the power supply. The development of solid-state batteries promises further improvements in safety and performance, potentially revolutionizing numerous applications within the renewable energy sector.

Similarly, fuel cell technology is undergoing significant development. The pursuit of cleaner and more efficient fuel sources, primarily hydrogen, necessitates parallel progress in hydrogen production, storage, and distribution infrastructure. The challenges are significant – developing cost-effective and safe methods for hydrogen production from renewable sources, building robust storage solutions, and creating a widespread distribution network are all critical factors. Overcoming these obstacles will unlock the vast potential of fuel cells in transportation, stationary power generation, and portable applications.

Furthermore, energy storage itself is not a monolithic entity. A diverse range of solutions exist, each with its own set of advantages and limitations. Pumped hydro storage offers large-scale capacity but is geographically restricted. Thermal energy storage excels in specific applications but suffers from inherent energy losses. Compressed air energy storage provides flexibility but faces challenges in efficiency and scalability. The choice of an appropriate storage solution depends critically on the specific application, the scale of the energy system, and the characteristics of the renewable energy source being integrated.

The scope of this field extends beyond the mere technological aspects. The economic implications of transitioning to a renewable energy-based system are profound, influencing investment strategies, energy pricing, and the overall economic competitiveness of nations. Furthermore, the environmental benefits are undeniable, with the potential to mitigate climate change and improve air quality. Policy decisions and regulatory frameworks play a crucial role in shaping the adoption and deployment of renewable energy technologies, necessitating careful consideration of the broader societal and political contexts. Finally, the success of the global transition to a sustainable energy future hinges on collaborative efforts across various sectors. Researchers, engineers, policymakers, and industries must work in tandem to overcome the technological, economic, and political hurdles that lie ahead. International cooperation is essential to accelerate innovation, share knowledge, and establish global standards that facilitate the widespread adoption of sustainable energy solutions. The journey towards a cleaner, more sustainable energy future requires a multifaceted, collaborative, and comprehensive approach. This book aims to provide a detailed understanding of the key components and challenges involved in this vital endeavor. The road ahead is paved with opportunities for innovation, growth, and a healthier planet.

1.2 Key Concepts and Terminology

Understanding the terminology is crucial for navigating the world of renewable energy. Photovoltaics, often shortened to PV, refers to the direct conversion of sunlight into electricity using semiconductor materials. These materials, primarily silicon, absorb photons, creating electron-hole pairs that generate an electrical current. The efficiency of this process depends on factors like material purity, cell design, and the intensity of sunlight. Different PV technologies exist, each with its strengths and weaknesses. Silicon-based cells dominate the market due to their high efficiency and relatively low cost, but thin-film solar cells are gaining ground due to their flexibility and potential for lower manufacturing costs. Concentrated solar power (CSP) systems utilize mirrors or lenses to focus sunlight onto a receiver, increasing the intensity and efficiency of solar energy conversion.

Fuel cells are electrochemical devices that convert the chemical energy of a fuel directly into electricity, in contrast to combustion engines. A key component is the electrolyte, a material that allows the movement of ions between the anode (where fuel is oxidized) and the cathode (where oxygen is reduced). Different types of fuel cells exist, distinguished by their electrolyte and fuel. Hydrogen fuel cells use hydrogen gas as the fuel, producing only water as a byproduct, making them exceptionally clean. Direct methanol fuel cells (DMFCs) offer a potential advantage in portability because methanol is liquid and easier to store than hydrogen gas. However, methanol-based fuel cells have lower efficiency than hydrogen fuel cells. The choice of fuel cell technology depends on the specific application, balancing factors like efficiency, cost, and availability of fuel.

Energy storage is paramount for managing the intermittent nature of renewable energy sources like solar and wind power. Without effective storage, fluctuations in renewable energy production can cause instability in the power grid. Battery technologies, such as lithium-ion batteries, are widely used for various applications, from electric vehicles to grid-scale energy storage. They store energy through electrochemical reactions, converting chemical energy into electrical energy and vice versa. Capacitors and supercapacitors offer faster charging and discharging rates than batteries, making them ideal for applications requiring quick bursts of power. These technologies differ in their energy density and power density characteristics. Large-scale energy storage is often achieved through methods like pumped hydro storage, which involves pumping water uphill during periods of low energy demand and releasing it to generate electricity during peak demand. Other large-scale options include thermal energy storage, which utilizes the heat capacity of materials to store energy, and compressed air energy storage, which compresses air at low demand and releases it to drive turbines at peak demand. The most suitable storage technology is dependent on factors such as scale, cost, and application requirements.

The efficiency of energy conversion and storage is significantly influenced by material properties. Nanomaterials, materials with features on the nanometer scale, offer exciting opportunities to improve the performance of renewable energy technologies. These materials can enhance light absorption in photovoltaic cells, improve the catalytic activity in fuel cells, and boost energy density in batteries. Additive manufacturing, also known as 3D printing, is a revolutionary technique that allows for the creation of complex components with precise control over shape and material properties. This technology is used to manufacture customized parts for solar cells, fuel cells, and other energy storage devices. The sustainability of manufacturing processes and material selection is also crucial. Choosing materials with low environmental impact and employing energy-efficient manufacturing techniques are essential for the long-term sustainability of renewable energy technologies. This holistic approach ensures that we reduce our carbon footprint and improve the overall lifecycle efficiency and sustainability of renewable energy systems. The integration of these technologies and materials requires careful consideration of their properties, limitations and practical applications. These are not isolated considerations but interconnected elements in creating a reliable, sustainable and cost-effective renewable energy infrastructure.

2 Photovoltaic Technology Fundamentals

Sunlight, the most abundant energy source on Earth, fuels photovoltaic (PV) technology. This technology converts sunlight directly into electricity using the photovoltaic effect, a phenomenon where light striking a semiconductor material generates an electrical current. Understanding this process is crucial for anyone interested in harnessing solar energy. The basic principle involves the absorption of photons, light particles, by the semiconductor material, exciting electrons to a higher energy level. This creates electron-hole pairs, which are then separated by an internal electric field within the PV cell, generating a direct current. The efficiency of this conversion depends on various factors, including the material's band gap, its ability to absorb light, and the effectiveness of charge separation.

Silicon, the most prevalent element in the Earth's crust, forms the backbone of most commercially available solar cells. Crystalline silicon solar cells, known for their high efficiency and relatively long lifespan, dominate the market. These cells are made by carefully purifying silicon, then shaping it into wafers, which are subsequently processed to create the necessary p-n junctions, essentially creating a built-in electric field which enables the generation of current. The manufacturing process is sophisticated and energy-intensive, making silicon-based cells relatively expensive despite their widespread use. However, continuous research and development efforts are constantly driving down production costs. Improvements in silicon purity, crystal growth techniques, and surface passivation strategies contribute to these efficiency gains.

In contrast to crystalline silicon, thin-film photovoltaics offer a compelling alternative. These solar cells use thin layers of semiconductor materials deposited on a substrate, resulting in a lighter and more flexible design. This flexibility makes them suitable for integrating into building materials, creating aesthetically pleasing and functional solar-powered roofs or facades. While typically less efficient than crystalline silicon cells, thin-film technologies often boast lower manufacturing costs, making them a more economically attractive option for large-scale deployment. Amorphous silicon, cadmium telluride, and copper indium gallium selenide (CIGS) are prominent thin-film PV materials, each with its own set of advantages and disadvantages concerning efficiency, stability, and manufacturing complexity. The ongoing exploration of new materials and improved deposition techniques promise to further enhance their performance characteristics and broaden their applications.

Concentrated solar power (CSP) systems represent a different approach to harnessing solar energy. Instead of directly converting sunlight into electricity, CSP technologies focus on concentrating sunlight onto a smaller receiver area using mirrors or lenses. This concentrated sunlight heats a working fluid, which then drives a conventional power generation system, such as a steam turbine. CSP systems are particularly advantageous in regions with high solar irradiance and abundant land areas. They can achieve higher efficiencies compared to standard PV systems and enable thermal energy storage, allowing for continuous power generation even after sunset. However, CSP plants require significant infrastructure investments, which can be a major barrier to widespread adoption. Different CSP system designs exist, employing parabolic troughs, power towers, or Fresnel reflectors, each with unique characteristics in terms of cost, efficiency, and scalability. Their environmental impact is also carefully evaluated, considering factors such as land use, water consumption, and material sourcing.

The selection of a photovoltaic technology depends on several factors, including the application, cost considerations, environmental impact, and availability of resources. While silicon-based solar cells currently dominate the market due to their established technology and high efficiency, the advancements in thin-film photovoltaics and CSP technologies make them increasingly competitive alternatives. Ongoing research and development efforts continue to improve the efficiency, cost-effectiveness, and sustainability of all PV technologies, paving the way for a more widespread adoption of solar energy in the future. Choosing the right PV technology is a critical decision, demanding a careful assessment of various technological, economic, and environmental parameters specific to the application.

2.1 Silicon-based Solar Cells

Silicon is the workhorse of the solar cell industry. Its abundance, relatively low cost, and well-understood properties have made it the dominant material for decades. However, the pursuit of higher efficiencies and lower manufacturing costs continues to drive innovation in silicon-based solar cell technology. We're not just talking about incremental improvements; we're witnessing a revolution in design and manufacturing processes.

Crystalline silicon solar cells, the most common type, are manufactured from highly purified silicon ingots. These ingots are then sliced into thin wafers, which undergo a series of processes to create the photovoltaic effect. This involves doping the silicon with specific impurities to create a p-n junction, allowing for the generation of electricity when sunlight strikes the cell. The efficiency of these cells is closely tied to the quality of the silicon, the precision of the doping process, and the design of the cell's surface. Modern crystalline silicon cells boast efficiencies exceeding 20%, a significant leap from the early days of solar technology.

Beyond the traditional crystalline silicon approach, there's a growing interest in alternative silicon-based solar cells. These include thin-film silicon solar cells and other emerging silicon-based designs. Thin-film cells, as their name suggests, use significantly less silicon material compared to their crystalline counterparts. This leads to lower manufacturing costs and greater flexibility in terms of application, allowing them to be integrated into various surfaces and products. They may not match the efficiency of crystalline silicon cells, but their cost-effectiveness makes them a compelling alternative, especially for large-scale deployments or specific applications where flexibility is crucial. Research continues to push the boundaries of thin-film technology, with new materials and processing techniques continually emerging.

Recent advances in silicon-based solar cell technology involve focusing on improving light trapping. By carefully designing the cell's surface texture and incorporating advanced anti-reflective coatings, manufacturers can enhance the amount of light absorbed by the silicon, leading to increased efficiency. These light-trapping techniques employ nanostructures and other advanced materials to manipulate the light's path within the cell, maximizing its interaction with the silicon and thus improving energy conversion. This is a critical area of research, as even minor improvements in light absorption can translate to significant gains in overall efficiency.

The manufacturing process itself has also undergone significant evolution. Advanced techniques like laser-induced doping and selective etching allow for finer control over the cell's structure and properties, resulting in higher efficiency and improved reliability. These improvements are not merely about optimizing existing processes; they represent a shift toward more precise and efficient manufacturing methods, significantly reducing production costs and environmental impact. Automation and innovative manufacturing strategies have also played an important role in the cost reduction of silicon-based solar cells.

Another major development is the integration of silicon solar cells with other technologies. For instance, combining silicon cells with other materials like perovskites creates tandem solar cells which offer the potential for efficiencies exceeding those achievable with silicon alone. These hybrid approaches exploit the complementary properties of different materials to capture a wider range of the solar spectrum, resulting in a significant boost in energy conversion. This synergistic approach is at the forefront of solar cell research, promising breakthroughs in efficiency and performance. The challenge lies in optimizing the interface between different materials to ensure seamless energy transfer and long-term stability.

Beyond the technological aspects, the economic viability of silicon-based solar cells is paramount. Continuous efforts to reduce manufacturing costs through process optimization and economies of scale are vital to making solar energy a truly competitive energy source. Significant progress has been made in this area, making silicon solar cells increasingly accessible and affordable. Furthermore, government incentives and supportive policies play a critical role in stimulating the growth of the solar energy industry and driving down prices. Long-term stability and durability are also important factors influencing the cost-effectiveness of silicon solar cells.

Finally, environmental considerations are intrinsic to the future of silicon-based solar cell technology. The industry is working towards minimizing the environmental footprint of silicon production and solar cell manufacturing. This involves reducing energy consumption, adopting sustainable materials, and developing efficient recycling processes. These environmental considerations are not peripheral; they are fundamental to the long-term sustainability of solar energy as a clean and viable energy solution. The future of silicon solar cells is bright, marked by ongoing innovations driven by a commitment to both efficiency and environmental responsibility.

2.2 Thin-Film Photovoltaics

Thin-film photovoltaics offer a compelling alternative to traditional silicon-based solar cells, presenting advantages in cost-effectiveness, flexibility, and diverse material options. Their thinness allows for lightweight and easily integrated applications, unlike the rigid and bulky nature of their silicon counterparts. This makes them ideal for building-integrated photovoltaics (BIPV), where solar cells are incorporated directly into building materials, seamlessly blending energy generation with architectural design. Imagine a future where roofs and facades generate electricity, reducing reliance on the grid and mitigating environmental impact. This isn't science fiction; it's the potential of thin-film technology.

Manufacturing processes for thin-film solar cells differ significantly from those employed in silicon production. Instead of large, single-crystal silicon wafers, thin-film technologies utilize techniques like sputtering, chemical vapor deposition, or screen printing to deposit thin layers of semiconductor material onto a substrate. This lowers the manufacturing cost, as it reduces the need for energy-intensive processes and expensive raw materials. Furthermore, the flexibility in substrate choice—glass, polymers, metals—opens up a wider range of applications and design possibilities. The substrate acts as a support layer for the active photovoltaic material, enhancing the cell's durability and handling.

Several types of thin-film solar cells exist, each with its unique characteristics and advantages. Amorphous silicon (a-Si) cells are known for their relatively low manufacturing cost and high light absorption, particularly useful for indoor applications. However, their efficiency is comparatively lower than crystalline silicon cells. Cadmium telluride (CdTe) cells, on the other hand, offer higher efficiency and better stability, with a growing market share due to their robust performance and cost competitiveness. Copper indium gallium selenide (CIGS) cells stand out for their high potential efficiency and tunable band gap, making them suitable for various applications and environmental conditions. Each type necessitates careful selection of materials and optimization of processing parameters to maximize efficiency.

The choice of material system profoundly impacts the overall performance and cost-effectiveness of a thin-film solar cell. For example, while CIGS cells show promising efficiency, the availability and cost of indium present a challenge. Similarly, the toxicity of cadmium in CdTe cells raises environmental concerns, necessitating careful manufacturing processes and responsible end-of-life management. The development and adoption of less toxic and more abundant materials are thus vital for the long-term success and sustainability of thin-film photovoltaics. Research into alternative materials and improved manufacturing techniques is crucial to overcome these limitations.

One key factor influencing the efficiency of thin-film solar cells is the quality and uniformity of the deposited layers. Defects in the thin films can significantly reduce the overall performance. Advanced techniques such as plasma-enhanced chemical vapor deposition (PECVD) are crucial to achieving high-quality, pinhole-free films that allow efficient charge collection and minimize energy losses. The precise control over layer thickness and composition is also essential for optimizing light absorption and carrier transport within the cell. Understanding and addressing these nuances is vital for producing high-performance thin-film solar cells.

Thin-film photovoltaics are not without their challenges. While their lower manufacturing cost and flexibility are significant advantages, their efficiency is generally lower than that of high-quality silicon solar cells, although substantial progress has been made in this area. Durability and long-term stability remain crucial considerations, particularly in harsh environmental conditions such as high temperatures or humidity. Research efforts are focused on enhancing the stability and lifespan of these devices to ensure their reliability over extended periods. This involves improving the materials' resistance to degradation and developing robust encapsulation techniques to protect the cells from environmental factors.

Integrating thin-film solar cells into existing infrastructure and building designs is an important step in widespread adoption. The flexibility of thin-film technology allows for integration into a wide range of applications, from building-integrated photovoltaics to flexible solar panels for consumer electronics. Innovative approaches are needed to seamlessly integrate these cells into building materials while maintaining their aesthetic appeal. Collaboration between material scientists, engineers, and architects is essential to create visually appealing and effective building-integrated photovoltaics.

The future of thin-film photovoltaics looks bright. Ongoing research and development efforts are continually improving their efficiency, durability, and cost-effectiveness. Innovative materials, advanced manufacturing techniques, and improved device architectures are poised to propel the performance of thin-film solar cells to new heights. This, coupled with a growing global demand for renewable energy sources, positions thin-film technology as a key player in meeting the world's energy needs sustainably. The development of sustainable energy solutions is paramount for a greener future, and thin-film photovoltaics are well positioned to contribute significantly to that future.

2.3 Concentrated Solar Power

Concentrated solar power (CSP) harnesses the sun's energy on a larger scale than traditional photovoltaic systems. Instead of directly converting sunlight into electricity, CSP uses mirrors or lenses to concentrate sunlight onto a receiver, generating high-temperature heat. This heat then drives a conventional power cycle, typically using steam turbines to generate electricity. This approach offers several advantages, including the ability to store thermal energy for later use, addressing the intermittency inherent in solar power.

CSP systems come in various configurations, each with unique characteristics. Parabolic trough collectors, for instance, utilize long, curved mirrors that focus sunlight onto a receiver tube containing a heat-transfer fluid. These systems are relatively mature and cost-effective, making them a viable option for large-scale power generation. Power tower systems, on the other hand, employ hundreds or even thousands of heliostats – individually controlled mirrors – that track the sun and reflect sunlight onto a central receiver atop a tower. This design allows for higher concentration ratios, leading to potentially higher efficiencies. Finally, linear Fresnel reflectors consist of arrays of long, narrow mirrors arranged to focus sunlight onto a linear receiver. This configuration offers a balance between the cost-effectiveness of parabolic troughs and the higher efficiency potential of power towers.

The choice of CSP technology depends on several factors, including geographical location, available land area, and the desired power output. For example, regions with abundant direct sunlight and large, flat expanses of land are well-suited for parabolic trough or linear Fresnel systems. Power towers, with their higher concentration ratios, may be preferred in locations with less intense sunlight or when maximizing efficiency is paramount, despite their higher upfront costs. Careful consideration of these factors is crucial for designing an economically and environmentally sustainable CSP plant.

Beyond the choice of technology, efficient energy storage is crucial for maximizing the value of CSP plants. Unlike photovoltaic systems, CSP offers inherent capabilities for thermal energy storage. Molten salt, for example, is a commonly used storage medium. The high-temperature heat generated during the day can be stored in molten salt tanks and used to generate electricity even after sunset, extending power generation beyond daylight hours and improving grid stability. The duration of thermal storage can be tailored to meet specific needs, providing flexibility in dispatching power and matching the daily electricity demand profile. This is a significant advantage over other renewable energy sources that lack such built-in storage capacity.

However, CSP systems also face certain challenges. The high temperatures involved necessitate the use of specialized materials that can withstand extreme heat and harsh weather conditions. Furthermore, land use requirements for large-scale CSP plants can be substantial, raising environmental concerns about habitat disruption and land degradation. Careful site selection and environmental impact assessments are vital for mitigating these risks. Additionally, the initial capital cost of CSP plants can be higher compared to other renewable energy technologies, although this is gradually decreasing with technological advancements and economies of scale. Finally, efficient operation and maintenance are crucial for ensuring the long-term performance and reliability of CSP systems. Regular inspection and timely repairs are essential to minimize downtime and maximize energy production.

Therefore, concentrated solar power offers a promising pathway towards sustainable energy generation. Its ability to store thermal energy, generate electricity at high efficiency, and provide dispatchable power makes it a valuable component of a diversified renewable energy portfolio. However, careful consideration of various technological and economic factors is essential for optimal system design, implementation, and operation. Continuous research and development efforts are crucial for overcoming remaining challenges and realizing the full potential of CSP technology in a future powered by renewable energy. The ongoing development of more efficient materials, improved designs, and innovative storage solutions are steadily advancing the capabilities and competitiveness of CSP, paving the way for broader adoption across the globe. Careful planning and attention to both technical and economic realities will determine the success of its integration into the future energy landscape.

3 Fuel Cell Principles and Applications

Hydrogen fuel cells, a cornerstone of clean energy technology, directly convert the chemical energy stored in hydrogen into electricity through a fascinating electrochemical process. This conversion avoids the combustion process, resulting in significantly reduced emissions compared to traditional fossil fuel-based power generation. At the heart of the fuel cell lies a membrane separating two electrodes—the anode and the cathode. Hydrogen gas is fed to the anode, where it undergoes oxidation, releasing electrons and protons. These electrons travel through an external circuit, generating a usable electric current, powering everything from vehicles to stationary power systems. Simultaneously, the protons migrate through the electrolyte membrane to the cathode, where they combine with oxygen (typically from the air) and the electrons returning from the external circuit, producing water as a byproduct. The simplicity of this process, coupled with its potential to deliver high efficiency and near-zero emissions, makes hydrogen fuel cells incredibly attractive. However, technological hurdles, such as cost-effective hydrogen production and storage, along with durability concerns regarding the fuel cell components, continue to present ongoing challenges to widespread adoption.

Direct methanol fuel cells (DMFCs) represent another compelling fuel cell technology, particularly suited to portable power applications. Unlike hydrogen fuel cells that require separate hydrogen storage, DMFCs use liquid methanol as fuel, offering convenient handling and higher energy density. Methanol is oxidized at the anode, releasing electrons and protons that are transported to the cathode, in a manner analogous to hydrogen fuel cells. Oxygen, typically from air, completes the electrochemical reaction at the cathode, producing electricity and water. The compact nature of DMFCs makes them ideal for powering smaller devices like laptops, mobile phones, and other portable electronics. Nevertheless, DMFCs face limitations. Methanol crossover, where methanol molecules permeate the electrolyte membrane and reach the cathode, reduces efficiency and lowers overall performance. Further research into enhancing membrane selectivity and improving catalyst performance is crucial for overcoming this critical limitation and realizing the full potential of DMFCs in portable power applications.

The applications of fuel cell technology extend far beyond portable power sources. Stationary fuel cell systems, often deployed in buildings or industrial settings, offer an efficient and reliable way to generate electricity while reducing reliance on the grid. Such systems can range from small-scale units for homes to larger systems for commercial buildings, providing decentralized and clean power generation. Furthermore, fuel cells are finding increasing use in the transportation sector, powering vehicles ranging from buses and cars to forklifts and other material handling equipment. Fuel cell vehicles offer significant environmental advantages compared to their internal combustion engine counterparts, emitting only water vapor, and often producing significantly lower levels of pollutants. The development of efficient and durable fuel cell stacks, coupled with advances in hydrogen storage and distribution infrastructure, are vital for widespread integration of fuel cell vehicles.

However, the path to widespread adoption of fuel cell technology is not without its challenges. The high cost of fuel cells remains a significant obstacle, although economies of scale and technological improvements are progressively reducing manufacturing costs. The need for efficient and safe hydrogen production and storage is equally important. Hydrogen, while abundant, is not readily available in its pure form; its production often requires significant energy input, potentially offsetting environmental benefits unless renewable sources are employed. Furthermore, establishing a robust hydrogen refueling

infrastructure presents a considerable logistical challenge, particularly in the transportation sector. Research is actively exploring various avenues for improving hydrogen production efficiency, such as electrolysis using renewable energy sources, and advancing the development of safe and reliable hydrogen storage solutions. The focus on developing cost-effective and durable fuel cell materials and improving overall system efficiency is also pivotal.
In summary, fuel cell technology represents a promising avenue towards a sustainable energy future. While challenges remain in terms of cost, efficiency, and infrastructure development, continuous innovation and research efforts are progressively overcoming these obstacles. The inherent advantages of fuel cells, including their high efficiency, clean operation, and versatility, are driving a growing global interest in their application across diverse sectors, from stationary power generation and portable devices to the transportation sector. The continued development and deployment of fuel cell technology hold the key to a cleaner and more sustainable energy landscape, offering an effective path towards decarbonizing energy production and consumption worldwide. Ultimately, the widespread adoption of fuel cells will depend on a multi-faceted approach that incorporates technological advancements, supportive government policies, and a concerted effort towards creating a robust hydrogen economy. The future of energy, to a significant extent, hinges on the successful navigation of these challenges and the effective integration of fuel cell technology into the global energy mix. The ongoing research and development efforts are crucial in achieving this ambitious, yet potentially transformative goal.

3.1 Hydrogen Fuel Cells

Hydrogen fuel cells represent a compelling pathway towards a cleaner energy future. They offer a high energy conversion efficiency, producing electricity through an electrochemical reaction between hydrogen and oxygen, with water as the only byproduct. This clean energy profile makes them attractive for various applications, from powering vehicles to providing electricity for remote locations. Their modularity allows for scalability, adapting to diverse energy demands, from small-scale portable devices to large-scale power plants. However, challenges remain in widespread adoption. The infrastructure required for hydrogen production, storage, and distribution is currently underdeveloped, creating a significant hurdle to overcome. Further research into more efficient and cost-effective hydrogen production methods is crucial for realizing the full potential of hydrogen fuel cells. The development of durable and affordable fuel cell components is equally vital. Addressing these technological and infrastructural challenges will unlock the immense possibilities presented by this promising technology.

Unlike combustion engines, hydrogen fuel cells operate silently and efficiently, generating power without producing harmful emissions. This quiet operation, combined with the lack of pollutants, makes them particularly suitable for urban environments where noise and air pollution are major concerns. Imagine a city powered by silent, emission-free vehicles and energy sources; hydrogen fuel cells are key to making that vision a reality. The efficiency of energy conversion in fuel cells surpasses that of many other technologies, making them a highly competitive option for many applications. The inherent scalability of the technology enables its adaptability to varying energy needs, from small-scale applications like portable electronics to larger scale installations like power plants.

The electrochemical process within a hydrogen fuel cell involves the splitting of hydrogen molecules into protons and electrons. Protons pass through a membrane, while electrons travel through an external circuit, creating an electric current. Oxygen reacts with the protons and electrons at the cathode, producing water. This straightforward process yields high-efficiency energy conversion. The efficiency of this process depends on several factors, including the type of membrane, the catalyst used, and the operating temperature. Optimizing these parameters is crucial for enhancing overall fuel cell performance. The design and materials science behind these cells are continuously improving, leading to greater efficiency and durability.

Research into hydrogen fuel cells is incredibly active. Scientists are exploring new materials and designs aimed at reducing costs, improving durability, and enhancing performance. One area of focus involves developing more efficient and cost-effective catalysts, typically platinum-based materials, to accelerate the electrochemical reactions. Research is also underway to develop alternative catalysts that are less expensive and more readily available. The development of more robust and durable membranes is critical to the long-term operation of the cells, as membrane degradation is a major cause of fuel cell failure. Improved membrane design and manufacturing processes will ultimately lead to fuel cells with extended lifetimes and reduced maintenance requirements.

Beyond technological advancements, the economic viability of hydrogen fuel cells depends heavily on the cost of hydrogen production. Currently, most hydrogen is produced from natural gas, a fossil fuel. However, there is increasing interest in developing green hydrogen production methods, utilizing renewable energy sources like solar and wind power to electrolyze water, splitting it into hydrogen and oxygen. Scaling up green hydrogen production is essential to creating a truly sustainable hydrogen economy. This large-scale transition will require substantial investment in renewable energy infrastructure, hydrogen production facilities, and the distribution networks needed to transport hydrogen to its end-users. Government policies and incentives will play a critical role in driving this transition.

The future of hydrogen fuel cells looks bright, especially considering the pressing need for clean and sustainable energy sources. As research continues to improve the efficiency, durability, and cost-effectiveness of hydrogen fuel cells, they are likely to play a significant role in meeting future energy needs. The transition to a hydrogen-based energy system will not be immediate; it requires coordinated efforts across various sectors, including energy production, transportation, and infrastructure development. The development of a robust hydrogen infrastructure, including production, storage, and distribution networks, is crucial for the widespread adoption of this technology. The potential environmental benefits are considerable, and this technology is positioned to play a crucial role in mitigating climate change. Addressing the technological and economic challenges surrounding hydrogen fuel cell technology is critical for realizing this potential.

3.2 Direct Methanol Fuel Cells

Direct methanol fuel cells (DMFCs) offer a compelling alternative to traditional energy sources, particularly in portable applications. Unlike hydrogen fuel cells, DMFCs utilize methanol, a readily available liquid fuel, eliminating the need for bulky and complex hydrogen storage systems. This simplicity translates to smaller, lighter, and more easily deployable power sources. The core principle behind DMFC operation revolves around the electrochemical oxidation of methanol at the anode and the reduction of oxygen at the cathode, generating a direct current. This process is catalyzed by electrocatalysts, typically platinum-based materials, which play a crucial role in optimizing efficiency and lifespan. The efficiency of DMFCs, however, is significantly impacted by methanol crossover, a phenomenon where methanol molecules diffuse through the membrane separating the anode and cathode, leading to a reduction in overall cell voltage and energy output.

Optimizing the performance of a DMFC requires careful consideration of several key parameters. The choice of membrane is critical; it must possess high proton conductivity to facilitate ion transport while simultaneously exhibiting low methanol permeability to minimize crossover. Researchers continue to explore advanced membrane materials, such as Nafion-based polymers and composite membranes, to achieve this delicate balance. Moreover, the design and fabrication of the electrodes are crucial. Porous electrode structures with high surface areas provide ample sites for the electrocatalytic reactions to occur, maximizing the efficiency of methanol oxidation and oxygen reduction. Precise control over electrode porosity and catalyst loading is essential for achieving optimal performance.

The catalyst itself, usually platinum nanoparticles dispersed on a high-surface-area support, is another area of intense research and development. Platinum is an exceptionally effective catalyst but is also expensive and scarce. Consequently, much effort is focused on developing cost-effective alternatives, including platinum alloys, transition metal oxides, and novel nanomaterials that can match or even surpass the catalytic activity of pure platinum while significantly reducing its consumption. This exploration involves meticulous material characterization and computational modeling to understand the intricate relationship between catalyst structure, electronic properties, and catalytic activity.

Beyond material science, innovative cell designs are being explored to improve DMFC efficiency and longevity. Strategies like optimizing flow field design, enhancing mass transport, and employing novel cell architectures are all contributing to advancements in the field. Improved flow field designs promote uniform distribution of reactants across the electrode surfaces, thereby increasing the uniformity of the electrochemical reactions and overall power output. Effective mass transport is crucial to supply the necessary reactants to the catalyst sites and remove reaction products, preventing performance degradation due to concentration polarization. The exploration of alternative cell architectures is driven by the necessity to improve durability and reduce the manufacturing complexity associated with DMFC fabrication.

The practical applications of DMFCs span a wide range, driven by their inherent advantages in portability and ease of fuel handling. They are particularly well-suited for portable electronic devices, such as laptops, mobile phones, and handheld communication equipment. Their compact size and relatively high energy density make them attractive alternatives to traditional batteries. Furthermore, DMFCs show promise for powering small-scale unmanned aerial vehicles (UAVs) and other portable instruments demanding lightweight and self-contained power sources. However, significant technological hurdles remain before DMFCs can achieve widespread commercial adoption. Challenges include improving their long-term stability, reducing the cost of manufacturing, and addressing issues of methanol toxicity and handling. Continued research and development are needed to overcome these limitations and unlock the full potential of DMFCs as a sustainable and efficient energy solution.

Despite these challenges, ongoing research efforts are steadily improving DMFC performance and addressing limitations. Advances in electrocatalysis, membrane technology, and cell design are paving the way for more efficient, durable, and cost-effective DMFCs. Researchers are exploring innovative approaches, such as the use of alternative fuels, improved electrode materials, and novel cell configurations, to further optimize DMFC performance. The integration of advanced diagnostic tools and modeling techniques allows for a more comprehensive understanding of the complex electrochemical processes within DMFCs, leading to targeted improvements in design and operation.

The future of DMFC technology is bright, with promising developments continuously emerging. Ongoing research focuses on improving the durability and lifespan of DMFC components and exploring new manufacturing methods to reduce costs. Researchers are also investigating new fuel sources and alternative electrode materials to enhance performance. The implementation of advanced characterization techniques and numerical modeling provides deeper insights into DMFC operation, contributing to optimized designs and enhanced efficiency. As DMFC technology continues to mature, we can anticipate its increasing integration into various applications, providing a reliable and efficient power source for diverse needs, ultimately contributing to a sustainable energy landscape.

The long-term viability of DMFC technology hinges on addressing several key challenges. Cost reduction remains a major hurdle, particularly concerning the use of precious metals such as platinum in the catalyst layer. Research into cost-effective alternatives is crucial for widespread commercial adoption. Furthermore, optimizing the stability and lifespan of the DMFC is crucial for long-term reliability. The degradation of components, especially the membrane and catalyst, under operating conditions remains a significant obstacle. Therefore, improved durability is essential to increase the overall lifespan and reduce the maintenance costs. Finally, the safety aspects related to the handling and storage of methanol need to be addressed adequately, implementing safeguards and procedures to mitigate potential risks associated with the fuel's inherent toxicity.

4 Energy Storage Mechanisms

Lithium-ion batteries currently dominate the market, boasting high energy density and relatively long lifecycles. Their widespread use in portable electronics and electric vehicles speaks volumes about their practicality. However, challenges remain, including concerns about raw material sourcing, safety issues related to thermal runaway, and the environmental impact of their production and disposal. Research into solid-state batteries offers a promising pathway towards enhancing safety and improving energy density, potentially revolutionizing electric vehicle range and lifespan. Solid-state batteries replace the flammable liquid electrolyte with a solid one, thereby mitigating fire risks. The development of novel solid electrolytes with high ionic conductivity and good mechanical properties is critical to their commercial viability. Ongoing research is focused on optimizing the manufacturing process and lowering costs to make them competitive with traditional lithium-ion batteries.

Capacitors and supercapacitors offer a different approach to energy storage, excelling in rapid charge and discharge capabilities. They store energy electrostatically, unlike batteries which rely on electrochemical reactions. This difference leads to superior power density, making them ideal for applications requiring quick bursts of energy, such as hybrid vehicles and power grid stabilization. Supercapacitors, with their significantly higher energy density compared to conventional capacitors, bridge the gap between batteries and traditional capacitors, offering a compelling blend of energy and power density. However, their energy density still lags behind that of batteries, limiting their applicability in situations requiring prolonged energy storage. Ongoing research focuses on developing new electrode materials and electrolytes to increase energy density while retaining high power capabilities.

Pumped hydro storage is a well-established, large-scale energy storage method. It involves pumping water uphill to a reservoir during periods of low energy demand, then releasing it to generate electricity during peak demand. This method offers considerable storage capacity and long lifespans, making it particularly suitable for grid-scale energy storage applications. The inherent scalability of pumped hydro allows it to integrate seamlessly into existing hydropower infrastructure or establish new facilities where geographic conditions are favorable. However, its suitability is limited by geographic constraints—requiring sufficient elevation differences and water resources—and the environmental impact of reservoir construction needs careful consideration. This often involves careful environmental assessments and mitigation strategies to minimize disturbance to ecosystems.

Thermal energy storage (TES) systems leverage the heat capacity of various materials to store thermal energy for later use. This can involve using molten salts, rocks, or even water to absorb heat during periods of surplus energy generation, subsequently releasing this heat to generate electricity or provide heating. TES offers a cost-effective solution for solar thermal power plants, smoothing out the intermittency of solar energy. Different TES technologies cater to various temperature ranges and applications, from low-temperature applications like building heating to high-temperature applications in industrial processes and concentrated solar power plants. The materials used in TES systems are crucial to its efficiency. For example, the melting point of the material will define the temperature range it can effectively store. Advanced materials engineering is leading to the development of more efficient thermal storage media.

Compressed air energy storage (CAES) uses excess energy to compress air, which is then stored in underground reservoirs or large tanks. Upon demand, the compressed air is released and used to drive a turbine, generating electricity. CAES offers a relatively inexpensive and scalable solution for long-duration energy storage, effectively handling the intermittent nature of renewable energy sources. This technology has the potential to balance electricity supply and demand over longer periods. However, efficiency losses during compression and expansion cycles need to be minimized. Research into advanced compressor and turbine technologies is crucial for improving the overall efficiency of CAES systems. Moreover, the potential environmental impacts, such as noise pollution and land use, need careful consideration during site selection and system design. Careful site selection is therefore a key aspect of making CAES a truly viable option.

4.1 Battery Technologies

Lithium-ion batteries currently dominate the market, boasting high energy density and a relatively long lifespan. Their widespread use in portable electronics and electric vehicles underscores their practicality. However, the reliance on lithium, a finite resource, presents a significant long-term challenge. Further research into improving lithium extraction methods and exploring alternative battery chemistries is crucial for sustainable energy storage solutions. Manufacturing processes also need continuous improvement to minimize environmental impact and maximize efficiency.

The quest for enhanced energy density continues to drive innovation in battery technology. Solid-state batteries, for instance, are gaining traction due to their inherent safety and potential for higher energy densities compared to their liquid electrolyte counterparts. This technology, while still under development, promises to revolutionize electric vehicle range and performance, addressing a key limitation of current lithium-ion batteries. The transition from liquid electrolytes to solid-state requires overcoming significant material science hurdles, however.

Beyond lithium-ion, several alternative battery chemistries are showing promise. Sodium-ion batteries, using the abundant element sodium, present a cost-effective and sustainable alternative. While their energy density is currently lower than lithium-ion, ongoing research is focused on bridging this gap. Likewise, research into magnesium-ion and zinc-ion batteries is active, exploring their potential advantages in terms of safety, cost, and environmental impact. These emerging technologies are poised to expand the battery landscape and provide tailored solutions for various applications.

Understanding the lifecycle of batteries is paramount. This includes not only their performance characteristics but also their manufacturing, use, and disposal. Recycling and reuse are increasingly crucial aspects of battery technology, reducing waste and minimizing the environmental footprint. Developing effective recycling techniques is critical to ensuring a sustainable future for battery-based energy storage. The industry is actively working on closed-loop systems that recover valuable materials, minimizing reliance on raw material mining.

The selection of an appropriate battery technology depends heavily on its intended application. For portable electronics, high energy density and long lifespan are key. Electric vehicles, however, often prioritize high power output and rapid charging capability alongside range. Grid-scale energy storage, on the other hand, requires very high capacities and long lifecycles, with cost-effectiveness often being a determining factor. Careful consideration of these factors is vital in optimizing battery performance and ensuring long-term viability. The choice, therefore, is context-dependent and demands a thorough understanding of specific requirements.

Beyond chemical composition, battery design also plays a pivotal role in overall performance. Innovations in cell architecture, electrode materials, and thermal management systems significantly impact energy density, power output, and safety. Research into advanced electrode structures, for example, aims to increase surface area and improve electron transfer, enhancing charging rates and overall efficiency. Sophisticated thermal management is equally crucial for optimal battery operation, especially under high-power demands and varying ambient conditions. Optimizing these factors is crucial for maximizing the efficacy of existing and emerging battery technologies.

Finally, the integration of batteries into broader energy systems presents unique challenges and opportunities. Effective energy management strategies are vital to ensure optimal charging and discharging cycles, minimizing degradation and maximizing lifespan. Intelligent control systems can help optimize energy flow, minimizing waste and maximizing efficiency. Furthermore, intelligent integration with renewable energy sources allows for more effective utilization of intermittent power generation, smoothing out fluctuations and ensuring grid stability. This coordination represents a crucial step towards a truly sustainable energy future.

4.2 Capacitors and Supercapacitors

Capacitors store electrical energy electrostatically, accumulating charge on conductive plates separated by an insulating dielectric. Think of them as tiny reservoirs, quickly filling and emptying. This contrasts with batteries, which store energy chemically through electrochemical reactions, a much slower process. The key difference lies in power delivery: capacitors excel at providing high power bursts for short durations, while batteries offer sustained energy release over longer periods. Their applications are vastly different due to these fundamental distinctions.

Supercapacitors, also known as ultracapacitors, bridge the gap between traditional capacitors and batteries. They possess higher energy density than conventional capacitors but lower than batteries, yet they also boast much higher power density than batteries. This unique combination makes them ideal for applications requiring both high power and energy storage, such as hybrid electric vehicles, power backup systems and portable electronics. The technology harnesses electrochemical double-layer capacitance, where ions accumulate at the interface between an electrode and an electrolyte, creating a large surface area for charge storage. Materials science plays a crucial role in optimizing this process, influencing both energy and power density.

One crucial aspect of capacitor design is the dielectric material. This insulator determines the capacitance value; a thicker dielectric reduces capacitance, while a higher dielectric constant increases it. Common dielectric materials include ceramic, plastic films, and electrolytic solutions. The choice of dielectric directly impacts the capacitor's operating voltage and temperature range, along with its overall performance and reliability. Understanding the properties of the dielectric is therefore paramount for selecting the correct capacitor for a specific application, whether it's a simple circuit or a sophisticated energy

storage system.
Selecting the right capacitor involves careful consideration of multiple factors. Capacitance, voltage rating, and tolerance are key parameters to specify. The desired size and physical characteristics also play a role, particularly in space-constrained applications. The operating temperature range must be compatible with the operating environment and the type of capacitor chosen. This is where the practical experience of an engineer comes into play. Selecting a capacitor that's oversized for the application may waste space and money, while choosing one that's too small can lead to component failure. Accurate calculation and selection are fundamental for reliable performance.
Supercapacitors, with their unique energy and power density profile, are especially promising for numerous applications. Hybrid and electric vehicles use them for regenerative braking, capturing kinetic energy and converting it to electrical energy which is then stored. This helps improve fuel efficiency and reduce emissions. In portable electronics, they provide quick bursts of power needed by devices that demand high energy instantly. Their ability to withstand a large number of charge-discharge cycles also makes them superior to batteries in many applications. These advantages are driving extensive research into developing improved materials and manufacturing techniques. Therefore, understanding their operational principles and potential benefits is key for those working with renewable energy technologies.
The manufacture of both capacitors and supercapacitors is a complex process. For traditional capacitors, this often involves depositing thin layers of conductive material onto a dielectric substrate. Electrolytic capacitors involve creating a thin oxide layer on an anode and immersing it in an electrolyte. This process needs precise control of parameters like temperature, pressure, and material composition to guarantee performance and reliability. Defects in the manufacturing process can lead to reduced capacitance, leakage current, or even component failure. Quality control plays a critical role in achieving consistent, reliable capacitor performance across the entire production run.
For supercapacitors, electrode material selection is crucial. Materials with high surface area and good electrical conductivity, like activated carbon, graphene,

and conducting polymers, are commonly used. The manufacturing process often involves combining these materials with a binder and a current collector. The resulting electrodes are then assembled into a device along with an electrolyte. Careful control over the porosity and morphology of the electrode materials is crucial for optimizing their performance. Optimizing these factors allows for improved energy and power densities, leading to better-performing devices that contribute to a more sustainable energy future.

The environmental impact of capacitor and supercapacitor manufacturing should also be a consideration. The selection of materials and the energy consumption of the production process have substantial environmental consequences. The use of less toxic materials and the development of more energy-efficient manufacturing processes are important aspects of creating environmentally friendly energy storage technologies. Sustainable practices need to be integral to the design and manufacturing of capacitors and supercapacitors to minimize their ecological footprint. The adoption of environmentally friendly manufacturing processes is essential for moving towards a greener future. As such, companies are increasingly incorporating aspects of the circular economy into their operations.

Future advancements in capacitor and supercapacitor technology are driven by the pursuit of even higher energy and power densities. This requires exploring new electrode materials and electrolytes. Research into advanced nanomaterials, such as carbon nanotubes and MXenes, offers the promise of significantly enhanced performance. Improved manufacturing techniques, like 3D printing, provide opportunities for creating more efficient and customized devices. These ongoing developments continually extend the scope of applications and improve the overall efficiency of energy storage systems. These breakthroughs are vital for powering advanced technologies and supporting the growing demand for sustainable energy solutions.

4.3 Pumped Hydro Storage

Pumped hydro storage (PHS) leverages the potential energy of water to store and release electricity. It's a mature technology, already deployed at a significant scale globally, offering a relatively cost-effective and long-duration storage solution crucial for balancing intermittent renewable energy sources like solar and wind. Essentially, PHS systems pump water uphill to a reservoir during periods of low electricity demand or surplus renewable energy generation. When demand increases, the water is released downhill through turbines, generating electricity.

This simple mechanism offers considerable advantages. Unlike batteries, which have limited lifecycles and complex recycling requirements, PHS systems boast exceptionally long operational lives, often exceeding 50 years, with minimal maintenance needs beyond routine inspections. Their longevity translates to significant cost savings over their lifespan. Moreover, PHS offers a remarkably high round-trip efficiency, often exceeding 70%, meaning a substantial portion of the energy used for pumping is recovered during electricity generation. This efficiency is significantly higher than many battery technologies, particularly for longer duration storage applications. The scale of PHS is also impressive; large-scale PHS plants can store gigawatt-hours of energy, providing grid-scale support for vast regions.

However, PHS is not without its limitations. The most significant constraint is the geographic requirement for suitable locations. A PHS plant necessitates two reservoirs at significantly different elevations, connected by a system of pipes and tunnels. This topographical limitation restricts the feasibility of PHS in many regions, particularly those lacking readily available valleys and hills. Furthermore, the environmental impact of PHS should be carefully considered. The construction of large reservoirs can disrupt ecosystems, alter water flow patterns, and impact local communities. Careful site selection, along with mitigation strategies, are essential to minimize adverse environmental effects. The significant capital investment required for PHS construction is another factor. Building a PHS plant involves substantial upfront costs associated with land acquisition, civil engineering works, and equipment procurement. This high capital expenditure can present a barrier to entry for smaller operators and developing countries.

Despite these challenges, pumped hydro storage remains a vital component of the sustainable energy transition. The ongoing development of advanced technologies, such as the use of more efficient pumps and turbines, is enhancing the effectiveness and economic viability of PHS. Innovations in system design, including the exploration of smaller-scale and modular PHS systems, are also addressing some of the logistical limitations. Moreover, the increasing integration of renewable energy sources, coupled with the need for flexible and reliable electricity grids, is driving greater demand for long-duration energy storage solutions. This heightened demand stimulates research and investment in PHS, further improving its efficiency and expanding its deployment possibilities. Careful consideration of environmental and social factors, combined with advancements in technology and cost optimization, will determine the future role of PHS in a sustainable energy future. The successful implementation of PHS hinges on effective policy support, proactive environmental management, and continued innovation.

Careful planning is crucial for mitigating the environmental impact. This involves conducting thorough environmental impact assessments to identify potential risks and develop strategies for minimizing negative effects. For instance, the selection of suitable reservoir locations should consider the potential disruption to local ecosystems and water resources. Mitigation measures may include habitat restoration, water management strategies, and community engagement to address concerns. Furthermore, the integration of PHS systems into the power grid requires careful coordination to ensure reliable and stable electricity supply. This includes developing advanced control systems capable of efficiently managing the flow of energy between renewable sources, PHS, and the grid. Incorporating PHS into the grid infrastructure is vital for enhancing grid stability and accommodating the fluctuating nature of renewable energy.

The economic aspects of PHS are complex and multifaceted. The high upfront capital costs require careful financial planning and investment strategies. Government incentives, such as tax credits or subsidies, can play a crucial role in making PHS economically viable, encouraging wider adoption. Moreover, life-cycle cost analyses need to be conducted to compare the total cost of ownership of PHS with other energy storage options. Factors to be considered include construction costs, operational expenses, maintenance needs, and the potential for revenue generation from electricity sales. A comprehensive economic evaluation is vital to support informed decision-making and secure investment for PHS projects. The ongoing research into cost-reduction strategies, along with the increasing demand for energy storage, offers optimism for enhancing the economic competitiveness of PHS in the long term. This ongoing effort in improving economics and design could potentially open the door to a vast number of new PHS facilities worldwide.

4.4 Thermal Energy Storage

Thermal energy storage (TES) systems offer a compelling solution to the intermittency challenges inherent in renewable energy sources like solar and wind power. Unlike batteries, which store energy chemically, TES systems leverage the capacity of materials to absorb and release heat. This heat can then be converted into electricity using conventional technologies, effectively bridging the gap between energy generation and demand. Crucially, this technology is scalable, adaptable to diverse applications, and environmentally friendly—making it a significant player in the sustainable energy transition.

The core of a TES system lies in the choice of storage medium. Various materials possess distinct thermal properties, influencing their suitability for different applications. Water, for example, is a readily available and cost-effective option, especially for large-scale systems like concentrated solar power (CSP) plants. However, its relatively low energy density might necessitate larger storage tanks compared to other materials. On the other hand, molten salts offer significantly higher energy density, enabling more compact storage solutions. Their high operating temperatures are particularly advantageous for high-efficiency power generation, though careful material selection and containment are essential due to their corrosive nature. Finally, phase-change materials (PCMs) undergo a solid-liquid phase transition at a specific temperature, offering a very high energy storage density within a narrow temperature range, making them suitable for specific niche applications like building heating and cooling.

The practicality of TES hinges heavily on the efficiency of heat transfer between the storage medium and the working fluid. Efficient heat exchangers are crucial for minimizing energy losses during charging and discharging cycles. The design of these exchangers must consider factors like the specific heat capacity and thermal conductivity of the storage medium and the working fluid, as well as the overall system's operating temperature and pressure. In large-scale applications, advanced designs using multiple heat exchangers in parallel might be required to manage the substantial heat fluxes involved. Optimizing heat transfer processes through innovative exchanger designs is vital for improving overall TES system performance and reducing operational costs.

Beyond the choice of storage medium and heat exchanger, the system's overall design is critical to its effectiveness. This includes factors like insulation, which plays a critical role in minimizing heat loss to the surroundings. High-quality insulation can significantly improve the overall efficiency and storage capacity of the system, especially over long periods. Furthermore, the containment structure is important; it must be robust enough to withstand the operating temperatures and pressures and needs to be designed to prevent leakage or contamination of the storage medium. Finally, the overall system architecture, encompassing the various components' integration and control, will impact the overall effectiveness and economics of the system.

Choosing the right TES technology demands a careful evaluation of several factors. The scale of the application, whether it is for large-scale power generation or localized heating, will dictate the suitable storage medium and capacity. The required operating temperature is another crucial factor, impacting material selection and system design considerations. For example, applications requiring high-temperature heat, like industrial processes, might necessitate molten salts or other high-temperature storage materials, while lower-temperature applications could use water or PCMs. Economic considerations play a significant role; the initial investment costs of different TES technologies vary significantly, and the long-term operational and maintenance costs must be carefully weighed. Therefore, a thorough cost-benefit analysis is essential to determine the most economically viable TES solution for any given application.

Moreover, the integration of TES into existing energy systems needs careful planning. This requires understanding the characteristics of the renewable energy source supplying the TES system, as well as the demands of the end-user. For instance, a CSP plant employing TES requires careful synchronization between the solar energy collection, the heat storage, and the electricity generation processes. Similarly, building heating systems employing TES need optimized control strategies that balance energy storage and consumption to maximize efficiency and minimize operational costs. The successful deployment of TES is contingent upon effective system integration and control strategies. Intelligent control systems can optimize charge and discharge cycles, improving system efficiency and lifespan. Ongoing research and development efforts aim to improve the performance and cost-effectiveness of TES technologies, exploring innovative materials, advanced system designs, and intelligent control strategies. The future of TES looks bright, with numerous opportunities for innovation and deployment in diverse applications.

4.5 Compressed Air Energy Storage

Compressed air energy storage (CAES) offers a compelling solution for balancing the intermittent nature of renewable energy sources like solar and wind power. Unlike batteries, CAES systems can store large amounts of energy for extended periods, making them ideal for grid-scale applications. The basic principle is straightforward: during periods of low energy demand or excess renewable energy generation, electricity is used to power compressors, which pump air into underground caverns or depleted gas fields. This compressed air, under high pressure, stores the potential energy. When electricity is needed, the compressed air is released, driving turbines connected to generators, thereby producing electricity. This process is remarkably efficient, with newer adiabatic systems significantly improving energy conversion rates compared to older isothermal designs. The potential for large-scale deployment, coupled with the mature technology base, positions CAES as a viable contender in the energy storage landscape.

The most significant advantage of CAES lies in its scalability and long duration storage capabilities. Unlike batteries, whose storage capacity is limited by their physical size and chemical properties, CAES systems can be scaled to meet substantial energy demands. Furthermore, CAES can store energy for hours or even days, effectively smoothing out the intermittency of renewable energy sources and ensuring a consistent electricity supply. This long duration storage is particularly crucial for grid stabilization and maintaining power during peak demand periods or unexpected outages. This capacity to handle large-scale energy storage makes it especially appealing to utility companies grappling with the integration of intermittent renewable energy.

However, CAES also faces certain challenges. The initial capital costs of constructing underground storage facilities can be substantial, requiring significant upfront investment. Finding suitable geological formations for air storage is another crucial consideration, with factors like rock permeability and structural integrity playing significant roles in site selection. Furthermore, the process of compressing air generates heat, which represents an energy loss in traditional isothermal systems. This is where adiabatic CAES steps in; by allowing the compressed air to cool naturally, significant energy efficiency improvements can be achieved.

The efficiency of CAES systems varies depending on the design and operating conditions. Isothermal CAES, the older technology, suffers from significant energy losses due to heat generation during compression. However, adiabatic CAES significantly mitigates this loss by using the heat generated during compression to preheat the air during expansion, leading to an improved round-trip efficiency. This technological advancement has made CAES much more competitive with other energy storage solutions. Ongoing research focuses on further optimization of CAES systems, including exploring innovative materials for better compression and expansion processes, and developing more efficient heat management strategies.

From a practical standpoint, CAES requires careful site selection and construction. The underground storage facilities need to be carefully designed and engineered to ensure structural integrity and prevent leaks. The geological characteristics of the chosen site are critical; suitable formations possess adequate volume and permeability for efficient air storage and retrieval. Moreover, the design must accommodate for the enormous pressures involved in CAES, necessitating robust materials and careful quality control during construction. This careful planning and engineering contributes to the higher upfront capital costs, which represent a barrier to widespread adoption.

Despite the challenges, advancements in CAES technology are continually improving its efficiency and cost-effectiveness. Researchers are exploring the use of advanced materials to create more durable and energy-efficient compressors and turbines. Furthermore, innovative techniques for heat recovery and management are being developed to further enhance the overall efficiency of the energy storage process. The integration of CAES into smart grids and microgrids is also an area of active research, with promising possibilities for optimizing energy management and enhancing grid stability. This continued development and ongoing innovation positions CAES as a viable technology for large-scale energy storage in a future increasingly reliant on renewable energy sources.

In conclusion, CAES presents a promising avenue for addressing the intermittency challenges associated with renewable energy. While the high initial capital costs and the need for suitable geological formations pose some hurdles, its scalability, long duration storage capacity, and improving efficiency make it a competitive option compared to other energy storage technologies. Ongoing research and development, coupled with the increasing demand for sustainable energy solutions, point towards a potentially significant role for CAES in the future energy landscape. The journey toward widespread CAES adoption will require further technological advancements, but the potential benefits are substantial. The ongoing innovation in the field is shaping the future of large-scale energy storage and its contribution to a more sustainable energy system.

5 Integrating Renewable Energy Sources

Successfully integrating renewable energy sources into our existing infrastructure requires careful planning and a multi-faceted approach. It's not simply a matter of installing solar panels; it's about creating a cohesive and efficient energy system. This demands a deep understanding of the interplay between different renewable technologies and the challenges they pose to the grid.

One key aspect is hybrid system design. Imagine a community powered by a combination of solar photovoltaic arrays, wind turbines, and a small-scale hydroelectric plant. Each source offers unique advantages: solar excels during sunny days, wind provides power even on cloudy days, and hydro offers a consistent baseload. However, effectively harnessing these requires sophisticated energy management systems to optimize energy production and consumption, accounting for the variability inherent in each renewable source. These systems typically employ smart inverters, energy storage solutions, and advanced control algorithms to balance supply and demand, ensuring a reliable power supply. Effective monitoring and data analysis are crucial to fine-tune these systems for peak efficiency.

The integration of renewable energy sources presents significant challenges to existing grids. Intermittency is a major hurdle. Unlike conventional power plants, solar and wind power fluctuate with weather conditions. This poses a threat to grid stability, potentially leading to voltage fluctuations and power outages. Addressing this requires a robust grid infrastructure capable of handling these fluctuations, often involving grid modernization projects that incorporate advanced sensors, smart meters, and communication networks. These technologies allow for real-time monitoring of power generation and consumption, enabling proactive adjustments to maintain grid stability. This also emphasizes the critical role of energy storage technologies, such as batteries or pumped hydro storage, in smoothing out these fluctuations and ensuring a reliable energy supply.

Furthermore, the integration of dispersed renewable energy sources necessitates improvements in grid infrastructure, particularly in terms of transmission and distribution networks. Consider the case of a remote wind farm supplying energy to a city. Efficient transmission lines are required to deliver the generated energy to its destination. This requires significant investments in upgrading existing infrastructure or building new transmission lines, which can be both costly and time-consuming. It also raises logistical and environmental considerations as these projects often involve navigating complex regulatory hurdles and minimizing environmental impacts. Careful route planning, environmental impact assessments, and community consultations are all crucial steps.

Beyond infrastructure, the successful integration of renewables necessitates advancements in energy management and distribution. Smart grids, leveraging advanced sensors, data analytics, and communication technologies, play a central role in optimizing energy flow and managing the unpredictable nature of renewable energy sources. These grids allow for dynamic adjustments to energy production and consumption, enabling more efficient use of resources and minimizing waste. Real-time data analysis aids in predicting energy demand and optimizing power generation, further enhancing efficiency and reliability. Smart meters also enable consumers to actively participate in energy management, potentially reducing their energy bills and contributing to a more sustainable energy future.

Effective grid integration also demands careful consideration of regulatory frameworks and market mechanisms. Policies and incentives can significantly influence the adoption of renewable energy technologies. Feed-in tariffs, renewable portfolio standards, and carbon pricing mechanisms can all encourage investment in renewable energy projects. However, these policies must be carefully designed to ensure fair competition and avoid unintended consequences. The integration process also requires transparent and reliable market mechanisms to facilitate the trading of renewable energy and integrate it seamlessly into the existing energy markets. Such mechanisms are crucial for ensuring a smooth transition to a sustainable energy system and attracting the necessary investments. Careful planning and collaboration between various stakeholders, including policymakers, utilities, and consumers, are essential for the successful implementation of appropriate regulations and market mechanisms.

Finally, the human element is equally critical. Successful integration requires skilled workforce development and public awareness campaigns to foster broader acceptance and understanding of the transition to renewable energy systems. This involves training professionals in areas such as renewable energy technologies, grid management, and energy efficiency, ensuring the availability of a skilled workforce to operate and maintain these systems. Similarly, public education campaigns are needed to highlight the benefits of renewable energy, address public concerns, and foster a sense of collective responsibility in achieving a sustainable energy future. A well-informed public is more likely to support the necessary changes and embrace the transition to cleaner and more sustainable energy systems.

5.1 Hybrid Systems Design

Designing effective hybrid systems requires a deep understanding of individual renewable energy components and their synergistic potential. We've explored photovoltaics, fuel cells, and various energy storage mechanisms separately; now, it's time to connect them strategically. Consider the complementary nature of solar power's peak output during sunny hours and the energy storage capacity of batteries, allowing for continuous power delivery even when the sun isn't shining. This integration is key to achieving reliable and sustainable energy solutions.

A well-designed hybrid system needs careful consideration of location-specific factors. For instance, a system in a region with abundant sunlight and limited wind resources might prioritize solar PV with battery backup. Conversely, a location with consistent wind but less sunlight might benefit from a wind turbine-fuel cell hybrid, using the fuel cell for continuous power generation and the wind turbine for primary energy generation. Detailed site assessments are essential. This involves analyzing local weather patterns, energy consumption profiles, and grid infrastructure to optimize system design.

Understanding load profiles is crucial for successful hybrid system integration. Residential loads, for example, typically exhibit peaks during the evening and lower consumption during the day. A system incorporating solar PV and battery storage can effectively manage this variation, storing excess solar energy during the day to meet evening demand. Similarly, industrial loads may demand constant power; a hybrid incorporating fuel cells alongside renewable energy sources would ensure continuous operation, even under conditions of reduced renewable energy generation. The type and size of the system must match the load profile of the facility.

Beyond the simple combination of technologies, smart control systems are vital for optimizing energy flow within the hybrid system. These systems monitor energy production from different sources, adjust power distribution based on demand, and manage energy storage efficiently. Algorithms are employed to prioritize energy sources based on factors such as cost, availability, and environmental impact. These systems ensure that the most efficient and economical strategy is employed at any given time, optimizing overall system performance.

The economic aspects of hybrid system design are equally significant. Initial capital costs must be balanced against long-term operational savings and potential revenue streams from energy sales or demand-side management programs. Life-cycle cost analysis, factoring in maintenance, replacement, and energy costs, is crucial for determining overall system profitability. Government incentives, such as tax credits and subsidies, may offset upfront investment costs, making hybrid systems more economically viable in certain regions.

Careful evaluation is essential for creating a financially sustainable project. Grid integration poses additional challenges for hybrid systems. Effective integration requires compliance with grid codes and standards, which vary across regions. This includes aspects like voltage regulation, frequency stability, and protection against grid disturbances. Interface equipment, such as inverters and power converters, play a critical role in ensuring seamless integration. Furthermore, the intermittent nature of renewable energy sources necessitates sophisticated grid management strategies to maintain overall grid stability and reliability. This often involves developing more flexible and responsive grid infrastructure and control systems, a critical area of ongoing research and development.

Finally, environmental considerations are paramount. The entire life cycle of a hybrid system, from material sourcing and manufacturing to operation and decommissioning, must be assessed for its environmental impact. Choosing environmentally friendly materials and manufacturing processes is crucial. The carbon footprint of the system, including greenhouse gas emissions associated with energy production and component lifecycle, should be minimized. Furthermore, proper disposal procedures are necessary at the end of the system's lifespan to prevent pollution. Sustainability must be the overall design principle. The choice of each component must be guided by minimizing the environmental footprint of the entire system.

5.2 Grid Integration Challenges

Integrating renewable energy sources like solar and wind power into existing electricity grids presents significant challenges. The intermittent nature of these resources—sunlight and wind are not consistently available—creates fluctuations in power supply, requiring sophisticated grid management strategies. This intermittency necessitates robust and flexible energy storage solutions to maintain grid stability and reliability. Without adequate storage, sudden drops in renewable energy generation can lead to power outages, impacting homes and businesses. This is a critical concern that requires immediate attention from engineers, policymakers, and energy companies alike.

Efficient grid integration relies heavily on advanced forecasting techniques. Accurate predictions of solar and wind power generation are essential for optimizing energy dispatch and preventing imbalances. This involves leveraging sophisticated weather models, real-time data from renewable energy facilities, and advanced algorithms to anticipate fluctuations in power supply. The accuracy of these forecasts directly impacts the effectiveness of grid management strategies and the overall reliability of the electricity system. Improved forecasting methods, therefore, remain a key area for ongoing research and development.

Beyond forecasting, managing the bidirectional power flows associated with distributed generation is equally vital. Unlike traditional power plants that primarily feed power into the grid, renewable energy sources can sometimes feed power back into the system, creating a two-way flow of electricity. This bidirectional flow requires upgrades to grid infrastructure and control systems to handle the dynamic nature of power generation and consumption. Smart grid technologies, incorporating advanced sensors, communication networks, and automated control systems, are essential for managing this complexity. This allows for more efficient use of resources and improved overall grid stability.

The integration of renewable energy also necessitates upgrades to transmission and distribution infrastructure. Existing grids, often designed for unidirectional power flows from large centralized power plants, may not be adequately equipped to handle the dispersed nature of renewable energy generation. Upgrading existing infrastructure, including transmission lines, substations, and distribution networks, is often necessary to ensure efficient and reliable power delivery. This can be a significant undertaking, requiring substantial investment and careful planning to avoid disruptions to the power supply. Investing in stronger, better placed and smarter infrastructure will be paramount going forward.

Furthermore, ensuring grid stability and reliability often requires the deployment of sophisticated control systems and strategies. These systems use real-time data to monitor grid conditions and make adjustments to maintain balance between power generation and consumption. This includes implementing fast-acting control mechanisms to respond to sudden changes in renewable energy generation and to prevent cascading failures. These control systems are becoming increasingly complex, requiring advanced algorithms and computing capabilities to manage the large volumes of data involved. Improving these systems is key to providing a steady and reliable source of electricity that's not dependent on fossil fuels.

Finally, the economic aspects of grid integration are crucial. The cost of upgrading grid infrastructure, implementing advanced control systems, and deploying energy storage solutions can be substantial. Careful economic analysis is needed to determine the optimal balance between investment costs and the benefits of increased renewable energy integration. This necessitates effective policies and incentives to encourage investments in grid modernization and to ensure a just transition to a sustainable energy future. The cost must not outweigh the benefit. A cost-effective transition is essential for widespread adoption.

Integrating renewable energy sources into the existing power grid presents a multitude of challenges, from technical complexities to economic considerations. The intermittency of renewable energy necessitates accurate forecasting, advanced grid management strategies, and robust energy storage solutions. Upgrades to grid infrastructure, sophisticated control systems, and effective economic policies are all essential for a successful transition to a more sustainable energy future. Addressing these challenges requires a collaborative effort among engineers, policymakers, and industry stakeholders to create a resilient and reliable grid capable of supporting a high penetration of renewable energy. The future depends on effective solutions to the challenges outlined here.

6 Advanced Materials and Manufacturing

The development of efficient and sustainable energy technologies hinges significantly on advancements in materials science and manufacturing processes. Consider, for instance, the ongoing quest for higher-efficiency solar cells. The pursuit of cheaper, more durable, and lighter materials is paramount. This requires a deep understanding of material properties at the nanoscale, leading to the development of novel nanomaterials with tailored characteristics. These advancements are not merely theoretical exercises; they directly translate into tangible improvements in the performance and longevity of photovoltaic devices, making renewable energy a more viable and cost-effective option for a wider population.

Nanomaterials, such as carbon nanotubes and graphene, offer exceptional electrical conductivity and mechanical strength, characteristics that can be exploited to improve the efficiency and durability of solar cells and fuel cells. The use of these materials allows for thinner and lighter devices, reducing material costs and simplifying manufacturing processes. Furthermore, the ability to tailor the surface properties of nanomaterials through controlled functionalization opens up avenues for enhanced light absorption and charge separation in photovoltaic devices. This level of precision contributes to maximizing energy conversion efficiency, thus increasing the overall sustainability and cost-effectiveness of the technology.

Additive manufacturing, more commonly known as 3D printing, is revolutionizing manufacturing across various industries, and its impact on energy technologies is undeniable. This technique allows for the fabrication of complex three-dimensional structures with intricate designs, impossible to achieve with conventional methods. For example, 3D printing can be used to create customized solar cell architectures, optimizing light trapping and charge carrier transport. This precision allows for the design of more efficient energy conversion systems. Beyond solar cells, additive manufacturing enables the creation of lightweight and highly porous structures for fuel cell components, enhancing gas diffusion and reducing material usage. This translates into decreased manufacturing costs and improved performance. Ultimately, it allows for quicker prototyping and iteration, accelerating the pace of innovation.

The selection of sustainable materials is crucial in minimizing the environmental footprint of energy technologies. Life cycle assessment (LCA) is becoming an increasingly important tool in evaluating the environmental impact of materials and manufacturing processes throughout the entire supply chain. This comprehensive approach considers factors such as energy consumption, greenhouse gas emissions, and waste generation during material extraction, processing, manufacturing, use, and disposal. By incorporating LCA into the design and manufacturing process, we can prioritize materials with low environmental impact and reduce the overall carbon footprint of the technology. This environmentally conscious approach promotes sustainable practices throughout the energy sector, ensuring that the transition to renewable energy is genuinely sustainable and responsible.

Moving beyond the immediate applications, the convergence of nanomaterials, additive manufacturing, and sustainable materials selection holds immense promise for future advancements in energy technologies. Imagine the possibilities: self-healing solar panels that repair minor damage, highly customizable fuel cells tailored to specific applications, and lightweight, energy-efficient energy storage solutions seamlessly integrated into buildings and infrastructure. The combination of these technologies allows us to surpass current limitations and create a more sustainable, efficient, and resilient energy future. Investing in research and development in these areas is not just an investment in technology; it's an investment in a cleaner and more secure future for generations to come. The path towards sustainable energy is paved with innovation and smart choices in material selection, pushing the boundaries of material science and manufacturing to build a truly lasting energy future. The successful integration of these advanced methods is key to unlocking the full potential of renewable energy and ensuring a sustainable future for all.

6.1 Nanomaterials for Energy

The unique properties of nanomaterials—their incredibly small size leading to high surface area to volume ratios—revolutionize energy technologies. This exceptional characteristic significantly enhances their performance in various applications, from improving solar cell efficiency to boosting battery capacity and enabling novel energy storage mechanisms. Think of it as dramatically increasing the interaction area available for chemical reactions or light absorption, resulting in much more efficient energy conversion.

Consider the use of carbon nanotubes (CNTs) in solar cells. Their exceptional conductivity and light absorption capabilities offer a significant improvement over traditional silicon-based cells. Incorporating CNTs can lead to increased power conversion efficiency and enhanced durability, extending the lifespan of solar panels. The lightweight nature of CNTs also makes them suitable for flexible and lightweight solar cells, opening up new possibilities for integration into various surfaces and applications. This is a far cry from the rigid, heavy panels of the past.

Nanostructured metal oxides, such as titanium dioxide (TiO2) and zinc oxide (ZnO), find widespread use in dye-sensitized solar cells (DSSCs). Their high surface area promotes better dye adsorption, leading to increased light harvesting and improved energy conversion efficiency. Furthermore, these materials are relatively inexpensive and environmentally benign, making them attractive for large-scale solar energy production. Their versatility is also considerable, with ongoing research exploring their use in other solar cell architectures.

Beyond solar energy, nanomaterials play a vital role in enhancing energy storage technologies. For instance, nanostructured materials like silicon nanoparticles are being explored as anodes in lithium-ion batteries. Their high surface area allows for faster lithium-ion intercalation and deintercalation, leading to improved charging and discharging rates. This translates to increased battery power and lifespan, crucial for applications such as electric vehicles and grid-scale energy storage. We are moving towards faster charging times and greater overall energy density.

Graphene, a single layer of carbon atoms arranged in a honeycomb lattice, possesses remarkable electrical conductivity and mechanical strength. Its use in supercapacitors can lead to significantly increased energy density and power density compared to traditional capacitors. This translates to the ability to store significantly more energy in a smaller volume and to deliver that energy much more rapidly. This is a game-changer for applications needing quick, high-power bursts.

Metal-organic frameworks (MOFs) are another exciting class of nanomaterials with potential applications in energy storage. Their highly porous structure allows for efficient gas adsorption, making them suitable for compressed air energy storage (CAES) systems. In CAES systems, compressed air is stored and then released to drive turbines for electricity generation. By enhancing the adsorption capacity and reducing pressure losses, MOFs improve the overall efficiency of CAES systems, making them more viable alternatives for large-scale energy storage.

The development and implementation of nanomaterials in energy technologies present several challenges. One major hurdle is the scalability of production processes. Manufacturing nanomaterials at an industrial scale requires careful consideration of costs and environmental impacts. Furthermore, long-term stability and safety concerns must also be addressed. The potential environmental effects of using nanoparticles needs thorough investigation and mitigating strategies developed. It's not simply about producing the materials, but also ensuring their safe and responsible use.

Despite these challenges, the potential benefits of nanomaterials in revolutionizing energy technologies are undeniable. Ongoing research and development efforts continue to push the boundaries of what's possible. Scientists and engineers are constantly exploring new ways to synthesize, characterize, and utilize nanomaterials to improve efficiency, durability, and cost-effectiveness across various energy applications. Their success will significantly contribute to our transition towards a sustainable energy future. From improving existing technologies to enabling entirely new ones, nanomaterials represent a crucial pillar of future energy innovations. The potential for a brighter, more sustainable energy future relies heavily on the continued progress in this field.

6.2 Additive Manufacturing Techniques

Additive manufacturing, also known as 3D printing, offers a revolutionary approach to creating components for renewable energy systems. Unlike traditional subtractive manufacturing, which involves removing material from a larger block, additive manufacturing builds objects layer by layer from a digital design. This allows for intricate designs and complex geometries previously impossible to achieve, opening up new avenues for innovation in solar cells, fuel cells, and energy storage devices. The process starts with a digital model, which is then sliced into numerous horizontal layers. These layers serve as instructions for the manufacturing process, guiding the deposition of material to create the three-dimensional object. This precision is crucial, particularly when dealing with the micro\item and nanoscale structures inherent in many advanced energy materials.
Consider the potential of 3D printing in creating customized solar cell designs. Instead of being limited to standard shapes and sizes, manufacturers can now tailor cell structures to optimize light absorption and charge carrier transport. This translates to higher efficiency and potentially lower production costs. Furthermore, 3D printing opens doors to the production of flexible and lightweight solar cells, ideal for integrating into various applications and environments, from wearable electronics to curved building surfaces. The ability to precisely control the composition and layering of different materials during the printing process further enhances the customization possibilities, enabling the creation of tandem solar cells with improved performance characteristics. The precise control allows for more efficient light trapping and charge collection, enhanc-

ing overall solar cell efficiency.

For fuel cells, additive manufacturing provides the opportunity to fabricate porous electrodes with enhanced surface area. These structures are vital for efficient electrochemical reactions within the fuel cell. The ability to design and print intricate three-dimensional pore networks maximizes the contact area between the reactants and the catalyst, significantly improving the fuel cell's performance and power output. Moreover, 3D printing allows for the integration of multiple components within a single manufacturing step, simplifying the assembly process and reducing costs. This streamlined approach to fuel cell fabrication holds the key to making fuel cell technology more affordable and widely accessible. The integration of different materials, such as catalysts and supports, becomes possible with precision, leading to optimized fuel cell designs.

In the realm of energy storage, additive manufacturing finds application in the fabrication of battery electrodes and other components. This precise manufacturing allows for the creation of electrodes with tailored pore structures, leading to improved ion transport and electrochemical reaction kinetics. The ability to precisely control the thickness and composition of the electrode layers allows for the optimization of battery performance characteristics, including energy density, power density, and cycle life. 3D printing is also proving useful in creating innovative battery architectures, such as three-dimensional electrodes, that overcome limitations inherent in traditional battery designs. These architectural innovations can significantly improve energy storage capacity and charging rates.

However, despite its immense potential, additive manufacturing in the energy sector faces certain challenges. One significant concern is material limitations. While the technology is rapidly advancing, not all materials suitable for energy applications are readily printable. Research efforts are ongoing to expand the range of printable materials and to develop printing methods compatible with these materials. Another challenge lies in scalability. While 3D printing is well-suited for creating prototypes and small-scale production, scaling up to mass production for commercial applications remains a hurdle. The speed of the printing process and the cost-effectiveness of scaling need to be addressed to make additive manufacturing a truly viable large-scale production method. Furthermore, ensuring the quality and consistency of the printed components is crucial. The layer-by-layer nature of additive manufacturing can introduce imperfections if not carefully controlled. Advanced quality control measures and process monitoring techniques are necessary to guarantee the reliability and long-term performance of the manufactured energy components. Stringent quality control is also critical to ensure consistent performance and durability, crucial aspects for the widespread adoption of the technology. Finally, the overall cost-effectiveness of additive manufacturing needs to be further improved, especially in relation to large-scale production.

Despite these challenges, the potential benefits of additive manufacturing in the renewable energy sector are substantial. As the technology continues to evolve and mature, it is poised to play an increasingly significant role in the development and production of advanced energy technologies, accelerating the transition towards a more sustainable energy future. Continuous improvements in material science and printing techniques will enhance the possibilities, enabling the creation of truly innovative and efficient energy solutions. The future of additive manufacturing in this sector looks bright, promising to revolutionize the way we design, manufacture, and implement renewable energy systems.

6.3 Sustainable Materials Selection

The environmental impact of material selection in renewable energy technologies cannot be overstated. Choosing sustainable materials is not just a matter of corporate social responsibility; it's a fundamental requirement for the long-term success and acceptance of these technologies. We must consider the entire lifecycle, from material extraction and processing to manufacturing, use, and eventual disposal or recycling. Ignoring this will undermine the very principles of sustainability we're striving for.

Consider the photovoltaic industry, for example. While silicon-based solar cells are currently dominant, the energy-intensive process of silicon purification has significant environmental consequences. This includes high energy consumption, greenhouse gas emissions, and the generation of hazardous waste. Therefore, exploring alternative materials with lower environmental footprints, like those based on perovskites or cadmium telluride, becomes crucial. Research into these materials is ongoing, focusing on increasing efficiency while minimizing environmental impact.

Manufacturing processes also demand careful consideration. The use of solvents and other chemicals in solar cell production can lead to air and water pollution. Therefore, the selection of manufacturing techniques that minimize waste and pollution is paramount. This includes implementing closed-loop systems for recycling solvents and exploring cleaner, more efficient production methods. Furthermore, the adoption of additive manufacturing techniques, such as 3D printing, offers promising possibilities to reduce material waste and customize designs for optimal efficiency.

Moving on to fuel cells, the choice of catalyst materials is significant. Platinum, a highly effective catalyst in many fuel cells, is a precious metal with limited availability and a high environmental cost associated with its extraction and refinement. Research is actively focused on developing alternative catalysts, such as those based on non-precious metals or metal oxides, that offer comparable performance with a reduced environmental footprint. This requires a multifaceted approach, incorporating materials science, electrochemistry, and nanotechnology.

Energy storage solutions, especially batteries, present a unique set of challenges. The extraction of lithium and cobalt, key components in many lithium-ion batteries, raises serious ethical and environmental concerns. These include habitat destruction, water pollution, and human rights issues in mining regions. Therefore, the search for sustainable alternatives, such as sodium-ion batteries or solid-state batteries, is gaining momentum. These alternative battery chemistries promise improved safety, longer lifespan, and a reduced reliance on environmentally problematic materials.

Beyond specific materials, the broader concept of circular economy principles must guide our choices. Designing products for recyclability and re-usability is essential. This necessitates careful consideration of material compatibility, ease of disassembly, and the development of efficient recycling processes. It is vital to move beyond the linear "take-make-dispose" model and embrace a circular approach that minimizes waste and maximizes resource utilization. This means working closely with material suppliers and developing strategies for end-of-life management of components.

Ultimately, the selection of sustainable materials for renewable energy technologies is a complex task requiring interdisciplinary collaboration. Engineers, material scientists, economists, and environmental scientists must work together to assess the full lifecycle environmental impacts of various materials and processes. It is crucial to develop comprehensive life cycle assessments (LCAs) for all components, taking into account all factors from extraction to disposal. This requires standardized methodologies and transparent data sharing. The information obtained will guide the development of greener technologies and supply chains. Only through careful and informed decision-making can we ensure that renewable energy technologies truly live up to their sustainability promises.

7 Future Trends and Challenges

The relentless pursuit of sustainable energy necessitates constant innovation. We've seen remarkable progress in photovoltaics, fuel cells, and energy storage, but significant hurdles remain. Overcoming these challenges will require a multifaceted approach, blending technological breakthroughs with sound policy and economic strategies. Imagine a world powered entirely by the sun, a world where our energy needs are met without sacrificing the planet's health—that's the ambitious yet achievable vision driving this research.

Artificial photosynthesis, a field brimming with potential, offers a pathway to directly convert sunlight into usable fuels, mimicking the natural processes of plants. While still in its nascent stages, the progress in developing efficient and cost-effective artificial photosynthetic systems is accelerating. Success in this area could revolutionize energy production, providing a clean and abundant energy source independent of intermittent solar or wind power.

Beyond artificial photosynthesis, advancements in energy storage are crucial. Current battery technologies, while improving, still face limitations in terms of energy density, lifespan, and cost. Solid-state batteries, for example, hold immense promise due to their improved safety and potential for higher energy density. Research into advanced materials and innovative designs could unlock superior energy storage capabilities, enabling the wider adoption of intermittent renewable energy sources. Furthermore, exploring alternative energy storage methods, such as improved pumped hydro systems or advanced compressed air energy storage, could address large-scale storage needs. The development and implementation of these technologies will depend heavily on further research and development as well as economic considerations.

However, technological advancements alone are insufficient. The transition to a sustainable energy future hinges on supportive policies and a favorable economic environment. Governments must implement policies that incentivize the development and deployment of renewable energy technologies, including tax credits, subsidies, and streamlined permitting processes. Carbon pricing mechanisms, while contentious, can play a crucial role in shifting market dynamics toward cleaner energy sources. Investment in research and development is also essential to further the advancement of these technologies and to bring down costs, making them accessible to a wider range of consumers and businesses. International collaboration will be crucial to overcome any hurdles. These incentives must be balanced with a consideration for the costs to the public, however, in order to ensure a fair and equitable transition.

Addressing workforce development is also critical. A skilled workforce is needed to design, build, maintain, and operate the next generation of renewable energy systems. Educational initiatives and training programs must be put in place to ensure a sufficient number of qualified professionals to meet the demands of a rapidly growing sector. This entails not only training in technical skills but also fostering an understanding of the broader economic and social implications of the energy transition.

Economic factors will heavily influence the adoption rates of these new technologies. The competitiveness of renewable energy technologies relative to fossil fuels depends on factors such as the cost of raw materials, manufacturing processes, and technological efficiency. Reductions in manufacturing costs and material supply issues must be addressed in order to enhance the price competitiveness of renewable energies in comparison to existing energy sources. The cost of energy storage also remains a significant barrier. Continuous technological advancements and economies of scale can help to lower costs, but government support may also be needed to further stimulate adoption.

The challenges ahead are significant, but not insurmountable. By combining technological innovation, supportive policies, and economic incentives, we can accelerate the transition to a sustainable energy future. The future of our planet depends on overcoming these challenges, and this process requires the cooperation of multiple stakeholders. While the task is demanding, the potential rewards—a cleaner, healthier planet powered by sustainable energy—are immeasurable. The path ahead necessitates a collaborative effort, bringing together scientists, engineers, policymakers, and the public to build a more sustainable energy future, one that secures a better world for generations to come. This is not merely a technological endeavor; it is a collective responsibility, a testament to human ingenuity and our commitment to a greener future.

7.1 Artificial Photosynthesis

Mimicking nature's elegant energy conversion process, artificial photosynthesis aims to replicate the way plants use sunlight, water, and carbon dioxide to create energy-rich molecules like glucose. This technology holds immense promise for a sustainable energy future, offering a pathway to directly convert solar energy into storable fuels. Unlike photovoltaic cells that produce electricity, artificial photosynthesis creates chemical fuels, solving the intermittency problem inherent in solar power.

The core of artificial photosynthesis lies in mimicking the intricate machinery of natural photosynthesis. Plants employ chlorophyll and a complex system of enzymes to efficiently capture sunlight and drive chemical reactions. Scientists are working to develop artificial systems using similar principles, albeit with synthetic materials and catalysts that are more robust, scalable, and adaptable to various environmental conditions. The challenge is immense; replicating the efficiency and sophistication of nature requires significant breakthroughs in materials science, chemistry, and nanotechnology.

One promising approach involves creating artificial leaves. These devices consist of a semiconductor material, typically a metal oxide or a hybrid organic-inorganic system, to absorb sunlight. This material generates electrons, which then drive redox reactions, similar to those occurring in natural photosynthesis. Water splitting is a key reaction where sunlight is used to break water molecules into hydrogen and oxygen. Hydrogen, a clean-burning fuel, can then be used in fuel cells to produce electricity on demand or stored for later use. The oxygen is a byproduct, released into the atmosphere.

Research is actively exploring a range of semiconductor materials and catalytic systems to improve the efficiency and durability of artificial leaves. Scientists are employing advanced characterization techniques to understand the mechanisms of charge transfer and catalytic activity, leading to more efficient designs. Nanostructuring materials can significantly enhance the surface area for light absorption and catalytic reactions. Optimizing the architecture of these artificial systems is crucial to maximize solar energy capture and minimize energy losses.

Beyond water splitting, researchers are investigating the possibility of directly converting carbon dioxide into useful fuels like methanol or methane using artificial photosynthesis. This approach could address climate change by utilizing atmospheric carbon dioxide, a major greenhouse gas. Integrating these artificial photosynthetic systems with other renewable energy technologies, such as wind and geothermal energy, could create a truly sustainable and resilient energy infrastructure. Further, the combination with existing energy storage solutions, like battery technologies, can provide the needed flexibility to address fluctuations in energy demand and supply.

However, significant hurdles remain before artificial photosynthesis can be deployed on a large scale. The efficiency of current artificial photosynthetic systems is still far below that of natural photosynthesis. The stability and cost-effectiveness of the materials and catalysts need to be improved. Developing scalable and cost-effective manufacturing processes is critical for wider adoption. Overcoming these challenges requires a concerted effort involving scientists, engineers, and policymakers. Investing in fundamental research and supporting the development of innovative technologies are essential for accelerating the progress in this field.

Furthermore, the integration of artificial photosynthesis into existing energy infrastructure presents a substantial challenge. The infrastructure, including pipelines and storage facilities, needs significant upgrades to accommodate the storage and distribution of hydrogen or other synthetic fuels. Policy support and economic incentives are vital to stimulate research, development, and commercialization of this technology. Public awareness and education are necessary to foster wider acceptance and drive innovation in this transformative area.

Successfully implementing artificial photosynthesis would be a significant leap forward in our transition to a sustainable energy future. It provides a pathway to abundant, clean, and storable energy, potentially mitigating the environmental impact of fossil fuels. While significant challenges remain, ongoing research and development efforts hold considerable promise. The potential benefits of this technology, coupled with continued innovation, suggest a bright future for sustainable energy production.

7.2 Energy Storage Advancements

Lithium-ion batteries currently dominate the market, offering a good balance of energy density and lifespan. However, their reliance on scarce materials and complex manufacturing processes presents challenges. Solid-state batteries, employing solid electrolytes instead of liquid ones, promise enhanced safety and higher energy densities. Research is actively pursuing solid electrolytes with improved ionic conductivity and stability, paving the way for more efficient and safer energy storage solutions. This improved safety profile is crucial, especially for large-scale deployments. Ongoing research is also focusing on reducing the production costs of solid-state batteries to enable wider adoption.

Beyond lithium-ion, other battery chemistries are emerging. Sodium-ion batteries, using readily available sodium, offer a cost-effective alternative, particularly suitable for stationary storage applications where energy density is less critical. Similarly, zinc-air batteries are being explored for their high theoretical energy density and low cost, though their practical lifespan and cycle life remain areas needing significant improvement. The development of new electrode materials and electrolytes is critical to improving the performance and overall competitiveness of these alternative technologies. We must also consider the environmental impact of mining and processing the materials needed for these batteries, pushing for more sustainable sourcing and recycling strategies.

Flow batteries offer a unique approach, separating energy storage from power delivery. They are particularly well-suited for large-scale grid storage due to their scalability and long cycle life. These systems use two liquid electrolytes that circulate between the electrode compartments, facilitating energy storage and release. The capacity is determined by the size of the electrolyte tanks, independent of the power output, providing flexibility in system design. Ongoing work centers on developing more efficient and cost-effective electrolytes, exploring different redox couples to optimize performance, and addressing issues like electrolyte degradation and system complexity. These solutions may also provide a pathway towards longer-lasting, grid-scale energy storage.

Beyond batteries, other energy storage technologies are advancing. Supercapacitors excel in rapid charge-discharge cycles, ideal for applications requiring short bursts of high power, such as hybrid electric vehicles and power grid stabilization. However, their energy density is generally lower compared to batteries. Advances are focused on increasing their energy density without compromising power density, such as exploring new electrode materials with enhanced surface area and improved electrolyte designs. This dual challenge requires considerable research efforts to reach a balance of high power and high energy.

Pumped hydro storage, a mature technology, remains a viable option for large-scale grid storage. It leverages the gravitational potential energy of water to store and release energy. However, the geographic limitations and environmental impacts of constructing large reservoirs restrict its widespread adoption. Improved efficiency through innovative pump-turbine technologies and exploring smaller-scale, decentralized pumped hydro systems are essential developments for this long-lasting approach. While well-established, it still offers viable solutions for grid stability.

Thermal energy storage (TES) provides a means to store energy as heat. Various TES systems exist, including sensible heat storage, latent heat storage, and thermochemical storage, each with unique advantages and disadvantages based on the temperature range and application. Improving the efficiency of thermal storage and developing innovative materials with high heat capacity and thermal conductivity is vital. This is crucial for improved heat transfer and reduced energy losses. The application of TES in conjunction with concentrated solar power plants is particularly promising.

Compressed air energy storage (CAES) offers another approach to large-scale grid storage. It involves compressing air during periods of low energy demand and using it to generate electricity when demand increases. Improvements in compressor and turbine technologies, along with addressing issues related to energy losses during compression and expansion, are key to enhancing the efficiency of CAES systems. Additionally, research is focusing on integrating CAES with other energy storage and renewable energy sources to create more effective and resilient energy systems. Its potential for grid-scale storage is high, but further research is necessary to improve overall efficiency and reduce costs.

The development and deployment of advanced energy storage technologies are fundamental to achieving a sustainable and reliable energy future. It's a multifaceted, ongoing global endeavor.

7.3 Policy and Economic Aspects

Government policies play a crucial role in shaping the renewable energy landscape. Incentives like tax credits, subsidies, and feed-in tariffs can significantly reduce the cost of renewable energy technologies, making them more competitive with fossil fuels. Conversely, a lack of supportive policies or the presence of policies that favor fossil fuels can hinder the growth of the renewable energy sector. Well-designed policies must consider the specific needs and contexts of different regions and countries, acknowledging the varied renewable resources available. For instance, a country with abundant solar radiation might benefit from policies prioritizing solar energy development, while another with substantial wind resources could focus on wind energy incentives. The effectiveness of these policies depends heavily on their design, implementation, and enforcement. Careful consideration of environmental impacts, economic feasibility, and social equity is paramount. Ignoring these aspects can lead to unintended negative consequences, such as environmental damage or social injustice.

The economic viability of renewable energy projects is equally important. The initial capital costs of many renewable energy technologies can be high, requiring substantial upfront investments. However, the operational costs are generally lower compared to fossil fuel-based power plants. This cost difference must be carefully analyzed over the lifetime of the project, considering factors such as maintenance, fuel costs, and potential revenue streams from electricity sales. Life-cycle cost assessments offer valuable tools for evaluating the overall economic performance of renewable energy projects. These evaluations must include not only direct costs but also indirect costs and external benefits, such as reduced greenhouse gas emissions and improved air quality. Furthermore, the economic benefits extend beyond the immediate project itself, often including job creation in manufacturing, installation, and maintenance of renewable energy systems. Therefore, a comprehensive economic assessment needs to incorporate regional economic impacts alongside the financial aspects of specific energy projects.

Successfully deploying renewable energy technologies requires a multifaceted approach. This includes not only technological advancements but also favorable policies and a robust economic framework. Government support, through various financial mechanisms, can significantly reduce financial barriers to renewable energy adoption. Additionally, creating stable and transparent regulatory environments encourages investment and development within the sector. Economic analysis is crucial for assessing the feasibility and profitability of renewable energy projects, while careful policy design can help to ensure that the benefits of renewable energy are widely shared. Technological innovation should also be fostered to further reduce costs and improve the efficiency of renewable energy technologies.

A balanced approach to energy policy is vital. This involves striking a balance between environmental protection, economic development, and energy security. Policies should strive to minimize environmental impacts while promoting the economic competitiveness of renewable energy. This balance is particularly crucial in transitioning from fossil fuels to renewable energy sources. Such transition necessitates careful planning and phased implementation to avoid disruptions in energy supply. Policies must also address issues of energy storage and grid integration to ensure reliable power supply even when renewable energy sources are intermittent. A well-defined regulatory framework for renewable energy integration is essential to manage the integration of these sources into the existing power grid safely and reliably. Failure to address these aspects can result in increased costs and instability in the energy sector.

Furthermore, international cooperation plays a crucial role in accelerating the global transition to renewable energy. Technology transfer, knowledge sharing, and joint research initiatives are all vital for promoting the widespread adoption of renewable energy technologies. Developing countries may require additional support, both financial and technological, to facilitate their transition. International agreements and policies can help to coordinate and harmonize efforts towards a global shift towards sustainable energy solutions. The establishment of standards and certifications for renewable energy technologies will enhance interoperability and promote fair competition in the market. This collaborative approach can foster innovation, accelerate the development and deployment of cost-effective renewable energy technologies, and eventually facilitate a worldwide transition to a sustainable energy future.

www.ingramcontent.com/pod-product-compliance
Lightning Source LLC
LaVergne TN
LVHW022027200926
848481LV00007B/2574